# The First-Timer's Guide to Science Fair Projects

By Q. L. Pearce and Francesca Rusackas
Illustrated by Kerry Manwaring

*Reviewed and endorsed by Anthony Fredericks, Ed.D.,*
*Associate Professor of Education, York College, York, PA,*
*and author of* The Complete Science Fair Handbook

LOWELL HOUSE JUVENILE

LOS ANGELES

CONTEMPORARY BOOKS

CHICAGO

*For David and Danny,*
*conductors of fun and love*
*—F.R.*

Publisher: Jack Artenstein
Director of Publishing Services: Rena Copperman
Executive Managing Editor, Juvenile: Brenda Pope-Ostrow
Editor in Chief, Juvenile: Amy Downing
Cover Photograph: Ann Bogart
Preparation of Science Fair Project on Cover: Susan Newman
Typesetting: Carolyn Wendt

Library of Congress Catalog Card Number: 97-74874

ISBN: 1-56565-734-9

Lowell House books can be purchased at special discounts when ordered in bulk for premiums and special sales.
Contact Department TC at the following address:

Lowell House Juvenile
2020 Avenue of the Stars, Suite 300
Los Angeles, CA 90067

Manufactured in the United States of America

10 9 8 7 6 5 4 3 2 1

# Contents

# To Parents and Teachers

A science fair is a fun and memorable experience. Your child (or student) will gain self-confidence through sharing with others what he or she has discovered. *The First-Timer's Guide to Science Fair Projects* takes your young scientist step by step through the thrill of developing and presenting a science fair project.

Each project in this book is safe and simple. The required materials are inexpensive and easy to obtain. Each project includes suggestions for research, complete with a list of key words that give a child direction when he or she begins to look up information. The "How They Work" section in the back of the book gives simple explanations of project results.

Help your child to find a project that interests him or her. With the Timetable on page 78, encourage your child to schedule plenty of time to do research, organize materials, and create an eye-catching and complete presentation. Promote creativity and organization. Resist directing your child's project but give plenty of support. With planning and patience, a science fair can be a marvelous way to uncover your youngster's gifts and talents.

**Key words to help give direction**　　**Type of project clearly labeled**　　**Creative suggestions for research**　　**Area of science being studied**

### Skin Deep
**(Demonstration/Experiment)**

**PROBLEM: Does the fat layer under the skin help keep people warm?**

**KEY WORDS**
fat, insulation, skin

**MATERIALS**
• scissors
• two empty 1-quart milk containers
• marking pen
• two indoor/outdoor thermometers
• spoon
• measuring cup
• shortening (room temperature)

**RESEARCH:** Check out a chart on the human body to see where fat layers are. Talk to a dermatologist about how the fat layer under the skin affects you. Interview someone at a local zoo or an aquarium about how animals rely on body fat. Find out how sea mammals stay warm in cold waters.

**HYPOTHESIS:** Record in your log book.

**PUT IT TO A TEST**

**1.** Cut off the top half of each milk carton. Label the milk cartons #1 and #2. Check that both thermometers show the same temperature, then spoon 1 cup of shortening into milk carton #1. Slip thermometer #1 about halfway into the shortening so that the bulb is completely covered but doesn't touch the sides or bottom of the milk carton. Be sure you can read the temperature.

40

**2.** Place thermometer #2 in empty carton #2. Put both cartons in the refrigerator side by side. Be sure to mark them "experiment in progress."*

**3.** Record the temperatures on both thermometers every 15 minutes for 60 minutes.

**4.** For a variation on this experiment, try placing both cartons in bowls of ice water instead of in the refrigerator.

**5.** Record what happened in your log book.

**CONCLUSION:** Record in your log book.

*It's always a good idea to carefully label any experiment you store in the refrigerator or in a kitchen cupboard. If you don't, your experiment may end up in someone's stomach!

**BRAIN BUSTER**

Fat helps humans to stay warm. What are some other ways the human body uses fat? Find out how the body stores its food energy.

**TIPS TO MAKE IT TOPS!**

A model cross section of human skin will make your display stand out. Use the Baker's Clay recipe on page 75 to form your model. Be sure to label each layer of skin, including the fat layer. Add details such as sweat glands and hair. By the way, did you know that early settlers used animal fat to make soap? The average human body stores enough fat to make seven bars of soap!

LIFE SCIENCE

41

**Inexpensive materials**　　**Simple steps**　　**Helpful illustrations**

# Your School Is Having a Science Fair!

So, now what? And what exactly is a science fair, anyway? A science fair is simply an event where you and your classmates display your latest science projects. In many fairs, teachers judge the projects and award prizes to winners. But even if you don't get a prize, you're a winner because you learn and have lots of fun taking part in the fair.

A science fair may be sponsored by a single class, or a whole city, county, or state. One of the first science fairs was the Children's Science Fair held in New York nearly sixty years ago. Now the National Science Fair is held every year to judge the best science projects in the country.

Do you like to do experiments? There are different categories of science fair projects, but most projects are experiments. In an experiment, you ask a scientific question, then design a test to find the answer. You can put together a project in other ways, too. You can demonstrate a scientific idea, present research, or show off a cool collection.

Every day, adventurous people make new scientific discoveries that affect your life. You can be a part of the adventure. By participating in a science fair, you get to think and work like a real scientist. The ideas in this book will help you get started. Change them, add details, make them your own. GOOD LUCK and HAVE FUN!

# Choosing a Project

Two very important steps must take place when you begin planning for a science fair. The first is to ask your teacher for a copy of the fair guidelines, so you know what rules and regulations to follow, and what restrictions there may be. The next step on your science fair adventure is choosing the right project. This is one of the most important steps you'll make, because you'll be working on your project for days, maybe even weeks. Begin by asking yourself these questions:

➤ What kind of science interests me?
➤ What would I like to learn about?
➤ What special hobbies or talents do I have that I could use to put together my project? (Examples are stargazing, bird-watching, or gardening.)

Write down several ideas that come to mind, and list the materials and equipment you will need for each. Then, to narrow down your choices, consider the following:

➤ Is the idea something you can easily study? (An investigation of the surface of Pluto might be cool, but it's not exactly practical!)
➤ Will you have to order some things through the mail?
➤ Do you have, or can you get, the money needed to buy supplies?
➤ Will you need any help with the project idea? From whom?
➤ Can you get the project ready by the due date?

If you're having trouble choosing an idea from your list, ask a parent or a teacher to help you select the best one. Once you have chosen the topic you'd like to work on, decide how you'll show it. Generally, you can do four things:

What could I do for my science fair project??

1. Present research.
2. Demonstrate a scientific idea or an apparatus.
3. Show a collection.
4. Perform an experiment.

## PRESENTING RESEARCH

Would you like to be a science detective? Then research is for you. For a research project, you ask a question, make observations, do plenty of reading, conduct interviews with experts, and use what you learn to discover a pattern or a trend.

For example, a good research question might be: "What type of seeds do wild sparrows like to eat best?" To find the answer, you could start by reading about sparrows at the library, then interview bird-watchers. You could also put out a variety of seeds and observe which birds come to the feeder (hopefully sparrows) and what type of seed they eat.

Another example of a research project is to test a product to find out if the manufacturer's claims are true (for instance, ask the question, "Which cereal really stays crispy in milk?").

## DOING A DEMONSTRATION

Would you like to demonstrate a scientific principle or fact, or show how a tool or a device works? Start by coming up with an interesting question, such as "How does rain form?" or "How is Gak® made?" or "How does a microscope work?" Then find a way to show the process. If you're demonstrating an apparatus, it's also fun to uncover the history of the equipment, how it is used, and why it is important.

Models and cross sections (objects cut in half to see inside) also make good demonstrations. For instance, you could show a model of the human eye or a cross section of a pair of roller blades.

## WHAT'S THE POINT?

For any science fair project you do, you need to explain the point, or *purpose*, of the project. For example, if you make a model of the human eye, your purpose needs to be clear. Is it to show how we see white light, or how we see color, or perhaps something else?

## SHOWING A COLLECTION

If you are a collector, this is your chance to shine. A display of a well-organized collection of rocks, shells, fossils, pressed leaves or flowers, insects, or anything else can be a great science fair project. To make your collection spectacular, show plenty of variety, and label everything clearly. Interesting pictures and maps of your discovery sites will jazz up your display. Also, if you've kept a log book detailing your discoveries, show that, too. (If you plan to come up with a collection in time for the science fair, keep a log book from the beginning. Read more about log books on page 13.)

## PERFORMING AN EXPERIMENT

The type of project most often presented at a science fair is an experiment. An experiment is a test that's designed to find the answer to a problem.

Exploring the universe through science is exciting, but there are rules to the game. To gather and present information in an orderly manner, scientists use the scientific method, a step-by-step approach to discovering answers and solving problems. In general, the steps are:

**1.** Find a problem. (Ask a scientific question that you are *able to test.*)
**2.** Do the research. (Gather as much information as possible.)
**3.** Make a hypothesis—that is, a guess. (Predict what the answer to the question will be. Be confident. Write your hypothesis in the form of a statement. Don't begin your statement with "I think.")
**4.** Experiment! (Think of a way to test your hypothesis. The test is the experiment.)
**5.** Record the results. (Collect data from the experiment.)
**6.** Draw a conclusion. (Figure out what the experimental data tell you by asking the questions below.)
   ➤ Do the results of your experiment tell you your hypothesis is on the right track or the wrong track? How?
   ➤ Is it possible to repeat the experiment? Should you change the experiment in any way?
   ➤ Did the experiment make you think of new questions that need answers?
   ➤ How can the information you found be useful? How does it relate to the world in which you live?

# "E" IS FOR EXCELLENT

Here are a few tips for doing a prize-winning experiment.

- Perform your test more than once to be sure your results are accurate. Repeat the first test exactly. (Each repeat is called a trial.) Record the results of each trial separately.
- Keep a log. For each trial, record the date and time, any measurements, observations, or results, as well as any comments you have.
- Be precise in taking and recording measurements and results.
- If possible, take photographs of noticeable changes that take place during the experiment.
- Be sure you don't gather only those results that say your hypothesis is correct. Finding the real answer is more important than proving your hypothesis is true.

## IMPORTANT FACTORS: VARIABLES AND CONTROLS

Your experiment should be carefully planned (that is, designed) to test one idea only. To do this, you will need a test *variable* and a *control* subject. A variable is a part (the one and only part) of the experiment that you change to test your idea. For example, to test whether plants need light, light should be the only variable.

Say you use four plants in your experiment. You begin by setting plant #1 aside, giving it enough light and water to keep it healthy and lush. Plant #1 is your control.

For the other three plants, the only thing you change is the amount of light each gets (more or less than the standard set for your control plant). The amount of light is your variable. Everything else, such as soil, water, air, type of plant, type of pot, and temperature, should be the same for all four test subjects.

## SAFETY FIRST

No matter what kind of experiment you do, it's very important to always follow a few safety rules:

➤ *Before* performing an experiment, plan it carefully with an adult. Decide together whether the adult should be there *during* the experiment. If the experiment calls for something electrical, hot, or sharp, an adult must be present.

➤ Know your tools and ingredients. Have each ready before you begin.

➤ Tie back long hair in a ponytail or pin it up.

➤ Keep your work area clean and dry. If necessary, cover surfaces with newspaper.

➤ Never put an unknown material in your mouth or eyes.

➤ Never use electrical appliances near water.

➤ Ask for help if something unexpected happens.

➤ Clean your work area when you're finished.

➤ Wash your hands after the experiment.

➤ Do no harm to a living thing.

# Doing the Project

Once you've decided on your project, it's time to make it a reality. A science project can be lots of fun, but there's plenty of work involved, too. That includes doing research, completing your experiment (or demonstration, collection, or research), writing your report, and building your display. Before you begin, find a calendar and clearly mark the due date of the project in red. Count how many days you have until that date, and use the Timetable on page 78 to plan each step.

## THE RESEARCH

No matter what kind of project you do, you need to gather as much information about your topic as possible. When doing your research, use a wide variety of resources, including:

- libraries (for books, magazines, cassette tapes, videotapes, and maps)
- government agencies
- local colleges and universities
- museums
- local science laboratories
- historical societies
- national parks
- hospitals
- planetariums
- zoos

You can get addresses and phone numbers for government agencies, historical societies, and national parks at your local library. Colleges, museums, hospitals, planetariums, and zoos are likely to be listed in the telephone book. When you call or write to these sources, ask if they have a public education department. They might have useful written materials they can send you. Call or write first to find out if they require a self-addressed envelope or, in some cases, a small fee.

Another way to do research (perhaps the most exciting way) is to contact an expert in the field you are studying and set up an interview.

Don't forget to keep a careful list of all your sources, no matter what they are. Make an index card for each source. You may need several cards for one source, depending on how much information you find. Here are some sample source cards to make:

## FOR BOOKS AND TAPES (include which it is on card)

Title:

Author:

Publisher:

Where published:

When published:

Page #:

Notes:

## FOR MAGAZINES AND PAMPHLETS (include which it is on card)

Name of magazine/pamphlet:

Title of article:

Author:

Date:

Volume:

Page #:

Notes:

## FOR INTERVIEWS

Person's name:

Person's title:

Person's workplace:

Telephone #:

Date of interview:

Notes:

## THE LOG BOOK

You will need to create a log book—an informal record you keep while your work is in progress. Your log book is the "story" of your project, containing your daily notes and thoughts. Judges often like to review log books to see how a student's work progressed.

The log book is an exception to the "neatness rule." It doesn't have to be retyped or rewritten. Drawings, smudges, and scribbled notes are okay.

## THE DISPLAY

The purpose of the display is to give a "project summary" at a glance. It is the first part of your science project that people will notice, so make it stand out. The display is made of tall boards. It should be easy to assemble and take apart, and sturdy enough to stand on its own for several days. The display board is usually two or three sections of strong material, sometimes held together with hinges. It can be made from pegboard, plywood, cork, particleboard, or foam board (cardboard and poster board are too flimsy to hold up over time). Many stationery supply stores carry lightweight, three-sectioned foam board, perfect for a science fair display. You'll have to check the rules to find out what size you'll need, but the average is about 3 feet by 5 feet.

You can cover your display board with fabric, burlap, or wallpaper to make it interesting. Over this board covering, you can put bright posters with clear, crisp lettering on them. You can also use snappy visual effects. For example, if your experiment is about plants, cut your display boards into large leaf or tree shapes. Use three-dimensional or exaggerated designs where possible. But be careful not to make your display so busy that people look only at it and not at your work!

Your display must include several things, each typed or lettered neatly on a separate poster (or on heavy paper).

1. A descriptive title of ten words or less
2. The purpose of your project (whether it's research, a collection, a demonstration, or an experiment)
3. Your hypothesis
4. A short summary of your procedures
5. A short summary of your results
6. A short summary of your conclusions

A clever title grabs attention. For example, "This Experiment Stinks!" would be a great title for your project if you did an experiment involving something smelly (see the project idea on page 38). The lettering should be easy to read and your title should be clear from a distance.

Use your space wisely. Fill the display board, but don't crowd things. Your presentation will be more spectacular if you use graphs, photographs, charts, drawings, diagrams, or samples. Triple-check your spelling and grammar, and remember that neatness counts.

## EXHIBIT MATERIALS

Think of the space in front of your display board as a scientific fun park. Show off experiment samples, cross sections, models, or tools. Remember, you won't always be standing at your project, so make it safe. If electricity is involved, nothing should get hot enough to burn someone or something. Place liquid or smelly samples in sealed, unbreakable containers. If any spillable material is unsafe, it is smarter to simply show photographs of it rather than have a sample. Be sure that electrical cords and small or sharp objects are out of reach of little children.

Most science fairs do not allow the use of live animals. However, that doesn't mean you can't study animals for your project. For example, if your project is a study of ant behavior, leave the critters at home and display photos only.

# THE WRITTEN REPORT

Your written report is a summary of your project. The judges will want to see a report that shows you understand the material you are presenting. Type it neatly and place it in a folder or binder. Decorate the cover of your binder or folder so that it goes with your display. Be sure to make it handy so anyone visiting the science fair can read it.

Your report needs to include:

➤ Title page (the title of your project, your name, your grade, and your school)
➤ Table of contents (a list of items included in the written report, with page numbers)
➤ The purpose of your project
➤ The problem being examined (if you do an experiment), as well as your hypothesis
➤ Research references
➤ List of materials used
➤ Procedures and observations
➤ Results
➤ Conclusion
➤ Acknowledgments (thanks to those who helped you)
➤ Any required forms
➤ Bibliography (a list of the resources you used to find information)

## HANDS OFF!

The following items are not allowed in science fairs:

• any body parts (except for teeth, hair, nails, or animal bone)
• hypodermic needles
• drugs
• dangerous chemicals
• materials that explode or catch on fire

You need to give proper credit to each kind of resource you use. In your bibliography, credit your resources as follows:

**For encyclopedias:** Name of encyclopedia, copyright date, topic, page number

**For books and videos:** Last name of author, first name of author, title of book or video, where published, publisher, when published

**For magazines:** Last name of author, first name of author, title of article, name of magazine, date of magazine, volume number, page number

**For interviews:** Person's name, person's job or position, person's workplace, address, phone number

# IN CASE OF EMERGENCY

It's helpful to bring along an emergency "fix-it" kit in case your project needs any last-minute repairing. Your kit should include tape, glue, scissors, a stapler, markers, and anything else you think you might need.

## THE JUDGE

A science fair judge looks for certain things. Originality is very important. Although it's okay to ask someone's opinion about your project, or to get adult help for your own safety, the judge wants to see that you did the project by yourself. Also, it's fine to follow a plan, but don't just copy an experiment from a book. Find a way to make the project uniquely yours. The following are some of the other things a judge looks for:

➤ Is the project complete?
➤ Is its purpose stated clearly?
➤ Is the information displayed creatively? Does it look good?
➤ Is the display safe?
➤ Is the display well-organized and clearly labeled?
➤ Is the written report neat and clear?
➤ Is the hypothesis stated clearly?
➤ Is the procedure explained clearly?
➤ Are the data easy to understand and do they support the conclusion?
➤ Have any variables been identified?

## YOUR TALK

In many science fairs, each student is given an opportunity to explain his or her project to the judge, and then answer questions about it. Don't worry! Judges are friendly and supportive, and they love science. You can use your display to guide them through the details of your work. If you don't know the answer to a question, it's okay. *A good project always raises new questions to be explored.*

On the day of your talk, dress neatly, be pleasant, and speak clearly. Finally, remember that your enthusiasm for your work can be contagious!

# Ideas for Science Fair Projects

In the following pages, you'll find many ideas for projects in the areas of physical science, life science, and earth science. The ideas cover the four major project types:

1. Research
2. Demonstration
3. Collection
4. Experiment

For any idea you choose, you'll want to mold and change it into something you can call your own. Also, be sure you understand the project fully. Look at the Key Words and Research sections to learn more about the subject.

You'll find that these projects are so much fun, you'll want to work on them whether or not you have a science fair in the near future!

# Balancing Act
### (Experiment)

## PROBLEM: How is the fulcrum important when using a lever?

### KEY WORDS
fulcrum, force, lever, load, simple machines

### MATERIALS
- marking pen
- tongue depressor (or craft stick)
- ruler
- pencil
- six or seven quarters

**RESEARCH:** At the library, research simple machines. Ask a parent to take you to a hardware store where you can look at a few different kinds of levers. With an adult, check out a construction site. Take pictures for your display.

**HYPOTHESIS:** Record in your log book.

## PUT IT TO A TEST

**1.** With a marking pen, label one end of the tongue depressor F for *force*. Label the other end L for *load*. Using the ruler, mark the center of the stick and label it B. Measure and mark the point halfway between F and B. Label that point A. Then measure and mark the point halfway between B and L. Label that C.

| F | A | B | C | L |
|---|---|---|---|---|

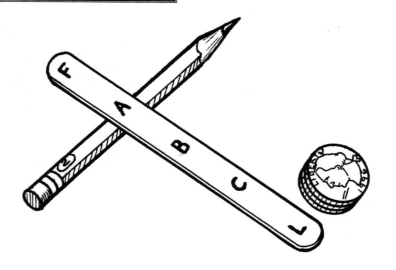

**2.** Place the pencil (your fulcrum) under the part of the stick labeled A. Place one quarter (your load) on top of the part labeled L. Place one quarter (your force) on top of the part labeled F. Does the force lift the load?

**3.** Place more quarters, one at a time, on top of the spot labeled F. How many does it take to lift the load? Carefully record your observations in your log book.

**4.** Remove all of the coins, move the fulcrum under point B, and repeat the process. Now how many quarters does it take to lift the load?

**5.** Remove all of the coins, move the fulcrum under point C, and repeat the process. Does it take more or fewer quarters to lift the load than it did in step 4? Does moving the fulcrum closer to the load make it easier to lift, or harder?

**6.** Record what happened in your log book.

 **CONCLUSION:** Record in your log book.

# BRAIN BUSTER

Using your model of a lever, find the point at which the force and the load are equally balanced. Where is that point? Record it in your log book.

TiPS TO MAKE iT TOPS!

For your presentation, collect and display pictures of levers used in everyday life. For example, a car jack, a teeter-totter, a crowbar, a wheelbarrow, a can opener, and scissors all use a lever in order to do their job. Even the human forearm acts as a type of lever. Label the lever and fulcrum in each example. Can you come up with a few more?

# Run It Up the Flagpole
### (Demonstration)

## PURPOSE: To show how a fixed pulley works.

## KEY WORDS
force, pulley,
simple machines,
work

**RESEARCH:** Go to the library and research simple machines. Ask your school principal if you can help to raise or lower your school flag. Ask an adult to take you to an auto repair shop. While there, ask a mechanic how pulleys are used to help lift a heavy load.

## PUT IT TO A TEST

**1.** Have an adult drill a hole at each end of the two yardsticks. The holes should be ½ inch from the ends and big enough for a pencil to fit through snugly.

## MATERIALS

- drill
- two wooden yardsticks
- one unsharpened pencil (thin enough to slip through the holes in the thread spools)
- small saw
- sandpaper
- two empty thread spools
- scissors
- construction paper
- marking pens
- 6 feet of light string
- adhesive tape

**2.** Ask your adult helper to saw the pencil in half. Sand down the jagged ends on each piece until they are smooth. Slip a pencil half through each thread spool. Make sure you can turn the spools easily.

**3.** Place the yardsticks side by side, so that the holes at their tops and bottoms line up. Keeping the spools on the pencils, slip the pencil ends through the holes in the yardsticks, creating a sandwich effect.

**4.** Cut out a small flag (about 4 inches by 6 inches) from construction paper and decorate it as you choose. Punch a *tiny* hole near the top and bottom corner of your flag. Thread the string through the holes. If the flag slips easily on the string, tape it in place.

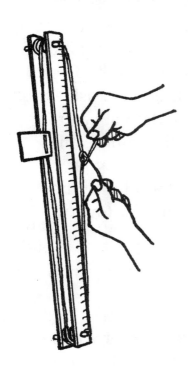

**5.** Loop one end of the string over the top spool and the other end over the bottom spool. Bring the ends of the strings together until they meet. Pull them tight and tie them together in a knot. The yardsticks, pencils, spools, and string form a fixed pulley. The two pulley wheels (the spools) are fixed—that is, they do not move up or down.

**6.** With your flag at the bottom of your "flagpole," pull the empty string down. In which direction does your flag move?

**7.** Record what happened in your log book.

 **CONCLUSION:** Record in your log book.

# BRAIN BUSTER

Using more than one pulley wheel spreads the weight of a load over more rope. That makes it even easier to lift such heavy loads as cars. Can you design a fixed pulley assembly using three or more wheels?

## TIPS TO MAKE IT TOPS!

**Window shades are usually raised by a small pulley. Find a way to include a pint-sized, homemade window shade as part of your display design.**

# Mix It Up
## (Demonstration)

## PURPOSE: To show the difference between a solution and a suspension.

### KEY WORDS

chemical,
element,
evaporate,
mixture,
sediment, solute,
solution, solvent,
suspension

### MATERIALS

• four small canning
  jars

• labels

• marking pen

• measuring cup and
  spoons

• hot tap water

• four stirring spoons

• sugar

• salt

• ground nutmeg

• ground allspice

**RESEARCH:** Go to the library and research solutions and suspensions. Call a local college or university and arrange an interview with an expert in chemistry. If you live near a body of water, such as an ocean, river, or lake, ask an adult to take you there. Then obtain a water sample so you can evaporate the sample and examine any residue left behind. Call your local water company and ask for a list of the elements or chemicals found in your town water supply.

## PUT IT TO A TEST

**1.** Label the first jar "#1—sugar." Pour ½ cup of hot tap water into the jar and stir in ½ teaspoon of sugar.

**2.** Label the second jar "#2—salt." Pour ½ cup of hot tap water into the jar and stir in ½ teaspoon of salt.

**3.** Label the third jar "#3—nutmeg." Pour ½ cup of hot tap water into the jar and stir in ½ teaspoon of nutmeg.

**4.** Label the fourth jar "#4—allspice." Pour ½ cup of hot tap water into the jar and stir in ½ teaspoon of allspice.

**5.** Observe the solution or suspension in each jar. Do the jars containing sugar and salt look the same as the jars containing nutmeg and allspice? Record your observations in your log book.

**6.** Wait 10 minutes. Now what do you see? Are the sugar and salt still in jars #1 and #2? Dip a clean finger in jar #1 and taste. Do the same with jar #2. How do jars #3 and #4 look?

**7.** Record what happened in your log book.

 **CONCLUSION:** Record in your log book.

## TIPS TO MAKE IT TOPS!

**Two solutions you made in this project are solids dissolved in a liquid. For your science fair project, make a list of other types of solutions or suspensions, such as:**

- **club soda—a gas dissolved in a liquid**
- **river or seawater—the sand is suspended in water**

## BRAIN BUSTER

You can go one step further with your project by observing what happens when you place jars #1 and #2 on a sunny windowsill and allow the sugar water and salt water to dry up.

Think about what other substances you can make into a solution or suspension in water. Are there some things that do not mix with water at all? Try a teaspoon of cornstarch, flour, or oil and observe what happens. Next try ½ teaspoon of cream of tartar and let it sit undisturbed overnight. What do you observe in the morning?

# Anchors Away

**(Experiment)**

## PROBLEM: What hull shape makes a boat go the fastest?

### KEY WORDS

boat design, hull, seamanship, sailing, surface tension

### MATERIALS

- aluminum foil
- glue
- three craft sticks
- three 4-inch-by-4-inch pieces of lightweight paper
- marking pen
- modeling clay
- scissors
- plastic wading pool (for infants) or large plastic tub
- stopwatch (or watch with a second hand)

**RESEARCH:** Go to the library and research different hull designs. Ask an adult to take you to a marina or a boat show to see various hulls up close. Arrange an interview with a boat owner who can help you with any questions you may have.

**HYPOTHESIS:** Record in your log book.

## PUT IT TO A TEST

**1.** Tear off three 6-inch-wide sheets of aluminum foil and fold each into a boat with a different hull shape. For example, make one flat on the bottom (like a raft), one rounded on the bottom, and one with a V-shaped bottom.

rounded    V-shaped    flat

**2.** Glue a craft stick to the center of each square of paper to make three sails. So you don't mix the sails up later, write a name or a number on each one.

**3.** Form three small, marble-sized balls of clay. Be sure they are all the same size.

**4.** Press a ball of clay into the inside center of each boat. Push the end of a craft stick (sail) into each one.

**5.** Outside, fill the wading pool with water. You can also use a pond or other small body of water if it is available. Place the first boat in the water near the edge, and mark the time you begin. You can do this experiment on a windy day by using a battery-powered fan or by blowing hard on the sail. Whatever you choose, try to make the amount of wind blowing on each boat the same. Try to keep the boat going straight, and time how long it takes it to get across to the other side. Repeat the experiment with each boat.

**6.** Record what happened in your log book.

 **CONCLUSION:** Record in your log book.

Are there other ways besides shape to overcome surface tension? Try sprinkling pepper across the surface of a small bowl of water. Then plop one droplet of dish soap right into the center of the bowl. What happens?

For your science fair project, you can also explore ways that nature puts surface tension to good use. For example, it allows tiny insects called pond skaters to actually walk on water!

## TIPS TO MAKE IT TOPS!

Since your project is about boats, come up with a sailing theme for your display. Blue and white background colors and a rope border around the display board can set the tone.

# Dueling Diapers

### (Research)

## PURPOSE: To discover which brand of disposable diapers holds more liquid.

## KEY WORDS

absorbency,
biodegradable,
capillary action,
diaper,
disposable

## MATERIALS

• samples of three different brands of disposable diapers

• marking pen

• paper towels

• food coloring

• measuring cup

• warm water

**RESEARCH:** Go to the library and look up diapers in a consumer guide or magazine. Talk to new parents and ask which type of diaper works best for their baby and why. Interview someone from a diaper service and a local environmental group to get different viewpoints on cloth versus disposable diapers.

## PUT IT TO A TEST

**1.** Label each diaper brand by name or number to help you keep track of each one. Then place diaper #1, plastic side down, on a paper towel.

**2.** Add several drops of food coloring to 1 cup of warm water. Slowly pour the water into the first diaper. Keep checking for

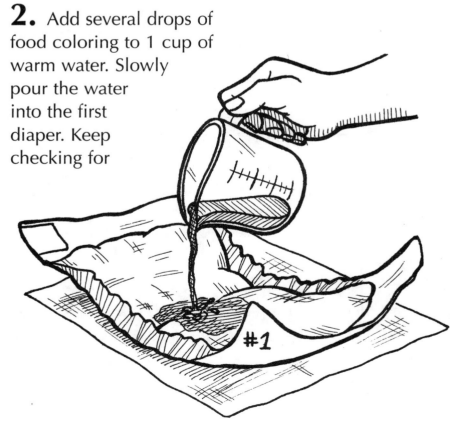

leaks on the paper towel. If you are able to pour the entire cup without a leak, prepare and pour more colored water. Make a note of how much the diaper holds before it leaks onto the towel.

**3.** Repeat the product test for diapers #2 and #3.

**4.** Record what happened in your log book.

 **CONCLUSION:** Record in your log book.

## BRAIN BUSTER

What are the advantages to using disposable diapers instead of cloth diapers? What are the disadvantages? Are any of the brands you test better for the environment (biodegradable) than others?

## TiPS TO MAKE iT TOPS!

Gather photographs and pictures from magazines of adorable babies wearing different kinds of diapers. Those cute little faces are sure to brighten up your display!

# Pucker Power!
## (Demonstration)

**PURPOSE:** To show how the first battery was made and how it works.

## KEY WORDS

battery, charge, chemical reaction, current, electricity, electron, proton, voltage

## MATERIALS

• scissors

• blotting paper (found at stationery stores)

• aluminum foil

• knife

• fresh lemon

• small bowl

• adhesive tape

• two 3-inch-long electrical wires with ends stripped ½ inch

• ten pennies

• LED (a light-emitting diode, available at hobby stores)

**RESEARCH:** Go to the library and research different kinds of batteries. Ask an adult to take you to a battery recycling factory, or interview someone from a battery manufacturing plant.

## PUT IT TO A TEST

**1.** To build a battery, cut ten penny-sized circles each from blotting paper and aluminum foil.

**2.** Ask an adult to cut the lemon in half and squeeze out all of the juice into a small bowl. Soak the blotting paper circles in the juice for about 30 seconds.

**3.** Tape one end of one electrical wire to an aluminum circle. With the wire on the bottom, place a juice-soaked circle of blotting paper on top of the aluminum circle. Place a penny on top of that.

**4.** On top of the first penny, build more layers: aluminum, more blotting paper, then a penny; aluminum, blotting paper, penny; and so forth. End with a penny.

28

**5.** Tape one end of the second piece of electrical wire to the top penny. You have now completed your battery.

**6.** Connect the loose ends of both wires to the LED. The LED should light up.

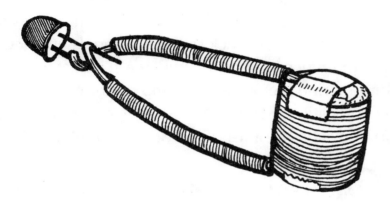

**7.** If you're still in the dark, make your battery taller (with more blotting paper, aluminum, and pennies), or substitute warm salt water or vinegar for lemon juice.

**8.** Record what happened in your log book.

 **CONCLUSION:** Record in your log book.

## A SHOCKING HISTORY

In the year 1800, an Italian scientist named Alessandro Volta discovered that he could produce a weak current of electricity by sandwiching together pieces of salt-water-soaked paper, silver, and zinc into a pile. As he touched a wire from the top of the pile to a wire from the bottom of the pile, sparks of electricity were generated. This was the first battery, and it is known as a voltaic pile. The electrical terms *volt* and *voltage* are named after Alessandro Volta.

## TiPS TO MAKE iT TOPS!

**Make a display of the various ways people use batteries. Include samples of different kinds of modern batteries.**

# Hello! Hello! Hello!

**(Experiment)**

## PROBLEM: What types of surfaces help to create an echo?

### KEY WORDS

acoustics, echo, echolocation, reverberation, sound waves

**RESEARCH:** Go to a library and look up echoes. Ask an adult to take you to a concert hall or a theater. Ask questions about the importance of acoustics in a theater. (For example, can echoes ruin a performance?) Check out the materials engineers use to reduce echoes. Interview a scientist at a zoo or a marine life preserve to find out how animals use echolocation. Sailors also use a type of echolocation to find schools of fish. Find out how it works.

**HYPOTHESIS:** Record in your log book.

## PUT IT TO A TEST

**1.** Mark an A on one end of tube #1, and a B on the other end. Lay tube #1 down on a flat surface, such as a desk or a tabletop. Slip the watch or the timer into the end of the tube marked A.

### MATERIALS

- marking pen
- two paper towel tubes
- watch or small timer that ticks softly
- book (about 8 inches by 10 inches)
- dish towel
- plate
- sweater
- sheet of aluminum foil
- flat rock

**2.** Stand the book upright at the end of the tube marked B as shown. Leave about a 1-inch gap between the tube and the book.

**3.** Place one end of tube #2 to your ear, then aim the other end toward any part of the book. What do you hear? Move the "ear tube" around to face any point on the book.

**4.** Now throw the dish towel over the book and repeat the experiment. Does the ticking get louder or softer? Take off the dish towel and carefully wrap a sheet of aluminum foil around the book, covering the area facing you. Is there a change in the echo? Remove the book. One by one, put the other materials in front of point B. What do you observe?

**5.** Record what happened in your log book.

 **CONCLUSION:** Record in your log book.

# BRAIN BUSTER

Many animals, such as bats, dolphins, and whales, use a method called echolocation to locate food. Find out how they use echolocation in this way.

# Magnetic Attraction
### (Experiment)

## PROBLEM: Can some nonmagnetic materials be magnetized?

### KEY WORDS

attraction,
compass,
electromagnet,
magnet,
magnetic field,
magnetic poles

### MATERIALS

- iron nail
- needle
- aluminum foil
- quarter
- steel screwdriver
- butter knife
- scissors
- rock
- paper
- bar magnet
- paper clips

**RESEARCH:** Go to the library and research magnets and electromagnets. Ask an adult to take you to a metal scrap yard to see an electromagnet at work. Interview someone who works in a stereo-speaker factory to find out how magnets are used there.

**HYPOTHESIS:** Record in your log book.

## PUT IT TO A TEST

**1.** Line up the following materials to magnetize: the nail, the needle, the piece of foil, the quarter, the screwdriver, the knife, the scissors, the rock, and the paper. Place the bar magnet near the paper clips, only to prove to yourself that they are attracted to it. Then remove the magnet.

**2.** Begin your experiment with the nail. Place it near the paper clips. Does the nail attract the clips? Now stroke the nail fifty times along one end of the bar magnet, always in one direction. Lift the magnet away from the nail after each stroke.

32

**3.** Test the nail with the paper clips once again. Does it attract the clips this time?

**4.** Repeat steps 2 and 3 to test each material. Which ones could be magnetized by the magnet?

**5.** Record what happened in your log book.

 **CONCLUSION:** Record in your log book.

**TiPS TO MAKE iT TOPS!**

**Instead of cutting plain squares of paper to display your written material, cut each piece in the shape of a bar magnet.**

# BRAIN BUSTER

Make your project more "attractive" by creating your own compass for display. Rub a bar magnet across a sewing needle at least thirty times in one direction. Fill a small bowl with water and place a drop of dish soap in the center. Float a quarter-sized slice of cork on the dish soap, then place the needle across the center of the cork (tape it if necessary). Spin the cork gently. When it stops, the needle will point north.

# Good Conduct

**(Experiment)**

## PROBLEM: Which materials are good conductors of electricity?

### KEY WORDS

circuit,
conductor,
electricity,
hydroelectric
dam, insulator,
terminal

### MATERIALS

- wire cutters
- plastic-coated electrical wire
- flashlight bulb screwed into free-standing socket-type bulb holder (available at hobby or hardware stores)
- screwdriver
- duct tape
- 1.5-volt battery
- aluminum foil
- sample materials (see step 5)

**RESEARCH:** Go to the library and research the ways people produce and use electricity. If you live near a hydroelectric dam, ask an adult to take you there. Talk to the people who work there and take pictures. Interview a local electrician. The electric company may be able to send you information that would be useful in your display.

**HYPOTHESIS:** Record in your log book.

## PUT IT TO A TEST

**1.** Ask an adult to help you cut three 3-inch lengths of electrical wire and strip ½ inch of plastic off each end.

**2.** Wrap one end of one wire around a terminal in the bulb holder. Screw the terminal in place. With the duct tape, tape the other end of the wire securely to either end of the battery.

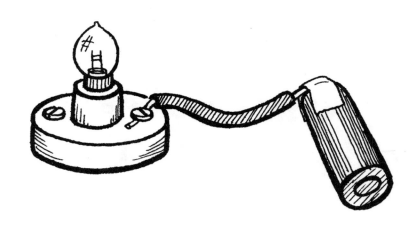

**3.** Tape one end of a second wire to the other end of the battery. Tightly wrap a pea-sized ball of aluminum foil around the loose end of this second wire.

**4.** Wrap one end of the third wire around the other bulb-holder terminal. Screw it in place. Tightly wrap the bare end of the third wire with a pea-sized ball of aluminum foil.

**5.** Line up your sample materials: a wood block, a quarter, a glass marble, a candle, a shell, a penny, and a butter knife. Pick up the two loose wire ends, one in each hand. (Hold the wires by the plastic coating, so the aluminum balls are free.) Touch the aluminum balls to each sample material. With what materials does the bulb light up? Try other materials, too.

# BRAIN BUSTER

Try this same experiment, but alternate links of paper clips, copper wire, and a pencil (sharpened at both ends). Will electricity flow through one of the conductors into another?

**6.** Record what happened in your log book.

 **CONCLUSION:** Record in your log book.

# Heat Stroke
## (Experiment)

## PROBLEM: Does the color of an object affect its temperature?

### KEY WORDS

absorption,
color, energy
conservation,
heat, reflection,
solar power,
temperature,
thermometer,
variable

**RESEARCH:** Interview an architect or a representative from the gas or electric company. Ask questions about the use of color in conserving energy. Interview a manufacturer or retailer of clothing. Ask how color affects body heat.

 **HYPOTHESIS:** Record in your log book.

## PUT IT TO A TEST

**1.** Paint the outside and bottom of three cups. Use a different color for each cup and let them dry. Even if your paper cup is already white, you still need to paint it white in order to keep your test materials the same. The thickness of the paint could be a variable.

**2.** Turn the cups upside down. Ask an adult to help you by cutting a small opening in the bottom of each cup just large enough to fit a thermometer snugly.

**3.** Read the temperature on each thermometer and record it. Then insert a thermometer, bulb first, a few inches into each cup. Lift the cups and check to see that the thermometer bulbs are not touching anything.

### MATERIALS

• white, green, and black water-based paint

• paintbrush

• four paper cups (nonwaxed)

• knife

• four indoor/outdoor thermometers

**4.** Place the three cups outside in direct sunlight. Be sure they are all on the same kind of surface. Place the fourth thermometer on top of the fourth, unpainted cup. That is your control. Check the temperature of all four thermometers every 10 minutes for 30 minutes. Record your findings.

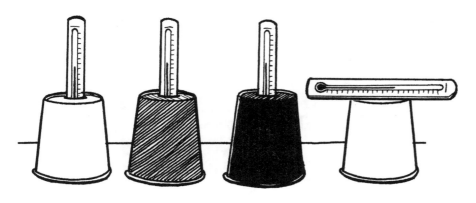

## BRAIN BUSTER

When your experiment is complete, move the cups and control thermometer indoors to a cool place. Once again, check the temperature of all four thermometers every 10 minutes for 30 minutes. Which cools down most quickly?

**5.** Record what happened in your log book.

 **CONCLUSION:** Record in your log book.

## TIPS TO MAKE IT TOPS!

Use the same colors on your display as you used for your cups. Create a large sun shape as a background for your title and a giant cardboard thermometer for your display.

# This Experiment Stinks!

**(Experiment)**

## PROBLEM: How can I detect acid in food?

### KEY WORDS

acid, base, chemical reaction, indicator, litmus paper

### MATERIALS

- one-quarter head of red cabbage
- knife
- large bowl
- hot water
- strainer
- large canning jar with lid
- labels
- marking pen
- plastic egg carton
- measuring spoons
- test materials (see step 3)
- mixing spoon

**RESEARCH:** Go to the library and research acids. Call a local college or a university and arrange to interview a chemist. Call a dietician and find out how acidic foods affect your health. Visit your dentist and ask how acids affect the health of your teeth and gums.

**HYPOTHESIS:** Record in your log book.

## PUT IT TO A TEST

**1.** Have an adult chop the cabbage into small pieces and place them in the bowl. Add enough hot water to cover the cabbage. Allow the cabbage to soak until the water is cool.

**2.** Strain the cabbage, collecting the cabbage water in the canning jar. Put the lid on the jar. Label the jar INDICATOR.

**3.** In each compartment of the egg carton, place about a tablespoon of each test material: lemon juice, milk, apple juice, smashed banana, and vinegar. Be sure to clean the measuring spoon after touching each material. Label each item.

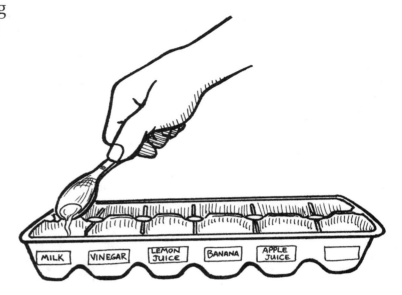

**4.** Carefully add a tablespoon of INDICATOR to each material and mix together with a mixing spoon. Again, carefully clean the spoon after touching each test material.

**5.** Check your INDICATOR for a color change in each egg-carton compartment. The INDICATOR turns red if the test material is acidic; it turns blue-green if the test material is a base (or basic). Write acid or base on each label.

**6.** Record what happened in your log book.

 **CONCLUSION:** Record in your log book.

## TIPS TO MAKE IT TOPS!

**To include a truly impressive demonstration in your science fair project, use a cotton swab dipped in lemon juice to write a secret message on a piece of typing paper. Let dry. Then have a judge use your cabbage INDICATOR to make your message appear!**

# Skin Deep
### (Demonstration/Experiment)

## PROBLEM: Does the fat layer under the skin help keep people warm?

**KEY WORDS**

fat, insulation, skin

**RESEARCH:** Check out a chart on the human body to see where fat layers are. Talk to a dermatologist about how the fat layer under the skin affects you. Interview someone at a local zoo or an aquarium about how animals rely on body fat. Find out how sea mammals stay warm in cold waters.

 **HYPOTHESIS:** Record in your log book.

## MATERIALS

- scissors
- two empty 1-quart milk containers
- marking pen
- two indoor/outdoor thermometers
- spoon
- measuring cup
- shortening (room temperature)

## PUT IT TO A TEST

**1.** Cut off the top half of each milk carton. Label the milk cartons #1 and #2. Check that both thermometers show the same temperature, then spoon 1 cup of shortening into milk carton #1. Slip thermometer #1 about halfway into the shortening so that the bulb is completely covered but doesn't touch the sides or bottom of the milk carton. Be sure you can read the temperature.

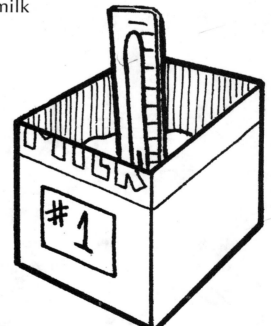

**2.** Place thermometer #2 in empty carton #2. Put both cartons in the refrigerator side by side. Be sure to mark them "experiment in progress."*

**3.** Record the temperatures on both thermometers every 15 minutes for 60 minutes.

**4.** For a variation on this experiment, try placing both cartons in bowls of ice water instead of in the refrigerator.

**5.** Record what happened in your log book.

 **CONCLUSION:** Record in your log book.

*It's always a good idea to carefully label any experiment you store in the refrigerator or in a kitchen cupboard. If you don't, your experiment may end up in someone's stomach!

# BRAIN BUSTER

Fat helps humans to stay warm. What are some other ways the human body uses fat? Find out how the body stores its food energy.

## TIPS TO MAKE IT TOPS!

A model cross section of human skin will make your display stand out. Use the Baker's Clay recipe on page 75 to form your model. Be sure to label each layer of skin, including the fat layer. Add details such as sweat glands and hair. By the way, did you know that early settlers used animal fat to make soap? The average human body stores enough fat to make seven bars of soap!

# Welcome to the Neighborhood
### (Research)

## PURPOSE: To create a Nature Guide of the main plants and animals in your neighborhood.

### KEY WORDS

animal, ecology, ecosystem, environment, habitat, native plant, plant

### MATERIALS

- local map
- backpack or knapsack
- pencils
- colored pencils or markers
- notebook
- camera
- plastic sandwich bags
- egg carton
- paper towels
- magnifying glass
- field guides (books) for plants and animals
- binoculars (optional)

**RESEARCH:** Interview members of local nature societies, such as bird, insect, or reptile societies. Ask what types of animals to look for in your area. Visit a local botanical garden to find out what plants are native to your area.

## PUT IT TO A TEST

**1.** Choose a target area of about one city block, or about one-half square mile in your neighborhood. Mark it on the map. If you live in a major city, you might focus your study on a nearby park.

**2.** Make up a field research schedule. Try to fit in at least two 30-minute observation periods per week for at least 4 weeks. Pack your supplies into an easy-to-carry backpack or knapsack.

**3.** With an adult partner, walk slowly around your target area and look for signs of wild animals, including tracks, nests, and burrows. When you find an animal to

include in your Nature Guide, take a photo or make a sketch of it. Observe the animal's behavior and take plenty of notes, such as what it eats, how it reacts to other animals, and what sort of call or sound it makes, if any. Don't forget to look for spiders and insects. These are animals, too!

**4.** You have plenty of plant neighbors as well, including trees. Photograph or sketch as many plant varieties as you can. Include pictures of flowers and seeds, if possible. Note the type of soil each plant grows in, and whether it prefers shade or sun. Do you see any insects feeding on the plants?

**5.** If you want to collect samples for your display, such as feathers, a leaf, pinecones, or a flower, be sure no creatures are living on them. *Do not* pull out a plant by the roots. *Do not* touch a plant you do not recognize. Carry your samples in plastic sandwich bags or in the compartments of an egg carton. Use field guides to help you identify animals and plants. Label everything, including all the drawings in your log book and all your samples.

# BRAIN BUSTER

You will often see different animals at night than you do in the daytime. For example, raccoons, opossums, bats, and owls are active at night. With an adult, plan three or four early-evening visits to your target area. Make a special "moonlight" section to go in your neighborhood Nature Guide.

**6.** Record your observations in your log book.

## TIPS TO MAKE IT TOPS!

For your display, make a garden-in-a-bottle! Gather seeds from small plants you find growing in the shade in your target area. Turn an empty, 2-liter plastic soft drink bottle on its side. Have an adult cut a 2-inch flap in the top that can be opened and closed. Through the flap, place small pebbles in the bottom, then add a 1-inch-deep layer of soil. Sprinkle in the seeds you have gathered and keep the soil moist. In 2 or 3 weeks or less, some of your seeds should sprout.

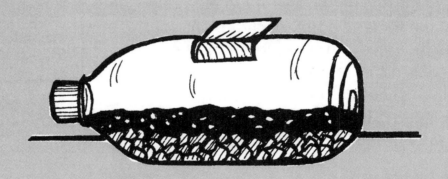

# Nature's Landfill
## (Experiment)

## PROBLEM: What materials will decompose naturally?

### KEY WORDS

bacteria,
biodegradable,
decomposition,
enzymes, litter,
organic, recycling

### MATERIALS

- ½-gallon milk carton
- scissors
- small picture frame glass (3 inches by 5 inches)
- ruler
- masking tape
- pie tin
- handful of pebbles
- garden soil
- assorted litter (such as apple slices, grass, Styrofoam cup, plastic candy wrapper, newspaper, sandwich bag, and paper bag)

**RESEARCH:** Go to the library and research landfills. Then visit a landfill. Talk with the operator about what items decompose faster than others. Talk with a local farmer about creating a compost pile. Ask an adult to take you to an arboretum. Ask workers there how they dispose of their dead leaves and plants. Interview a biochemist at a laboratory or a university. Visit a local recycling center.

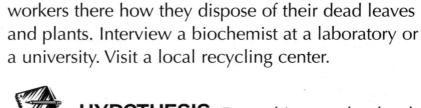

**HYPOTHESIS:** Record in your log book.

## PUT IT TO A TEST

**1.** Wash out the milk carton and open its top all the way, as shown. Have an adult cut an opening in the side of the carton. Make the opening slightly smaller than the picture frame glass, and about 3 inches from the carton top. Punch a few holes in the bottom of the carton for drainage.

**2.** Tape the glass to the inside of the milk carton, creating a window out of the opening cut in step 1.

**3.** Place the milk carton inside a pie tin. Put pebbles in the milk carton, then fill the carton up to the bottom of the window with soil.

**4.** Cut or tear each kind of litter into small pieces. Spread a thin layer of one kind of litter in the milk carton, then cover it with about ½ inch of soil. Spread a thin layer of a second kind of litter, and cover that with ½ inch of soil. Continue to make layers of different kinds of litter, then soil. Record the different materials in order of their appearance. Make sure you can see the layers through your window. End with a final layer of soil. Put a little water in the carton and let it filter down. Do not soak the soil.

# BRAIN BUSTER

Some plastic products are biodegradable. That means that they break down (degrade) when they are buried in a landfill. Test this for yourself. Ask an adult for permission, then dig two holes 6 inches deep in a garden. Bury a regular plastic sandwich bag in one, and a biodegradable sandwich bag in the other. With a waterproof marker, write "bio" on one craft stick and "regular" on the other. Mark each hole with the correct stick. Keep the area damp for 4 weeks, then dig up the bags and see if they have changed at all.

**5.** Close the top of the carton and put it in a warm, dark place. Keep the soil slightly moist. Over the next 3 or 4 weeks, observe the litter through the window.

**6.** Each week, record what happened in your log book.

**7.** Open the milk carton and remove some of each type of litter. Try not to disturb the sample directly in front of the "window." Note the changes that have occurred. Did everything deteriorate at the same rate? Be sure to save some "before" and "after" samples to display with your project.

 **CONCLUSION:** Record in your log book.

TIPS TO MAKE IT TOPS!

**Reuse some trash (aluminum cans, Styrofoam cups, and so forth) by making it into artwork! You can display your artwork as part of your science fair project.**

# Take a Little Off the Top
## (Experiment)

## PROBLEM: Does grass grow back no matter how short it is cut?

### KEY WORDS
grass, growth point, leaves, phototropism, stems

**RESEARCH:** Go to the library and look up grass to find out about its growth patterns. Interview a gardener or someone from a local nursery. Find a box of grass seed and write to the manufacturer printed on the box. Ask if the company has information it could send you.

**HYPOTHESIS:** Record in your log book.

### MATERIALS
- three plastic margarine containers with several small holes in the bottom
- potting soil
- measuring spoons
- rye grass seed
- misting bottle filled with water
- marking pen
- ruler
- scissors

## PUT IT TO A TEST

**1.** Fill one margarine container with soil, nearly to the rim. Sprinkle about a teaspoon of grass seed evenly across the top. Cover the seed with slightly more soil, and mist it lightly until it is moist.

**2.** Prepare the other two containers in the same manner. Label all three containers.

**3.** Place all three containers in a warm, bright place. Keep the soil moist and allow the grass to sprout and grow for 1 week.

**4.** Measure and record the average length of the grass in each container. Using scissors, cut the grass in container #1 by half. Cut the grass in container #2 to within ¼ inch of the soil. Cut the grass in container #3 as even with the soil as possible.

**5.** Continue to mist the grass for another week. Observe and record the growth in all three containers. Then repeat step 4. Do a third and fourth trial of your experiment.

**6.** Record what happened in your log book.

 **CONCLUSION:** Record in your log book.

# BRAIN BUSTER

Find the growth point for other fast-growing plants, such as alfalfa and radish.

# You've Been Fingered
## (Demonstration)

## PURPOSE: To show how fingerprints can be "lifted" from other materials.

### KEY WORDS

crime scene, criminologist, Federal Bureau of Investigation (FBI), fingerprints

**RESEARCH:** Go to the library and research fingerprints. Ask an adult to take you to a local police department. Ask police officers how they find and save fingerprints from a crime scene. Interview a detective or someone from a police science lab.

### MATERIALS

- various test materials (such as a glass jar, a doorknob, paper, an apple, and a desktop)

- talcum powder

- soft ¼-inch paintbrush

- clear cellophane tape

- black shiny paper (from a craft store)

## PUT IT TO A TEST

**1.** Don't wash your hands for a couple of hours, then touch each of your test materials in different places.

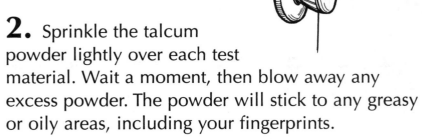

**2.** Sprinkle the talcum powder lightly over each test material. Wait a moment, then blow away any excess powder. The powder will stick to any greasy or oily areas, including your fingerprints.

**3.** Using the soft paintbrush, lightly brush the powdered areas to reveal the fingerprints. Take your time. You don't want to damage the prints. (You might have to practice this step a few times.)

**4.** Press a piece of clear tape over each fingerprint. Then carefully lift each one by pulling up the tape and sticking it onto the shiny black paper. You will see the powdered pattern of the fingerprint. Label what test material each print was lifted from.

**5.** Record your results in your log book.

The process of detecting and lifting prints was created by Sir Francis Galton in the mid-nineteenth century. Today, fingerprints give police a helping hand in fighting crime, and computers store the fingerprints of many criminals. Find out how such law enforcement agencies as the FBI use fingerprints to identify a criminal.

TIPS TO MAKE IT TOPS!

Somewhere in your display, be sure to mention that people are not the only ones to have one-of-a-kind prints. Cows have one-of-a-kind nose prints! When an owner enters a cow into certain contests, the owner must send in the cow's nose print. When it's time to pick the best heifer, the judge takes the contestant's nose print and compares it to the one on file. This prevents cheating through "cow switching." It's a moo-velous process, wouldn't you say?

# Red Sky at Night
## (Demonstration)

## PURPOSE: To show why the sky becomes red at sunset.

### KEY WORDS
atmosphere,
spectrum,
sunlight

### MATERIALS

- rectangular clear glass or plastic container (a small aquarium works well)
- water
- marking pen
- clay
- flashlight with fresh batteries
- measuring cup
- milk
- spoon

**RESEARCH:** Go to the library and research sunlight. Visit a local planetarium to see a model of how the sun's light strikes the earth. Interview a meteorologist or a sailor about how the color of the sky can tell you what weather may be coming.

## PUT IT TO A TEST

**1.** Put the container on a flat surface, then fill it nearly to the top with tap water. Mark one of the short sides A, and the other short side B. Mark the long sides C and D.

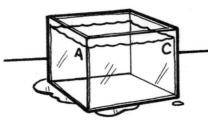

**2.** Make a ball of clay about the size of a baseball. Flatten one side of the ball a little so that it won't roll. Place it on the flat surface, about 3 inches from the edge of side A of the container.

**3.** Press the flashlight into the clay so that it stays in place. Adjust its position so that its light will shine right through the container. Be sure you are able to reach the switch easily.

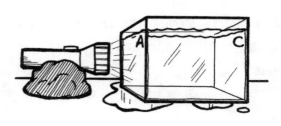

**4.** Pour ¼ cup of milk into the water and stir with the spoon. Allow the mixture to become still, then turn on the light.

**5.** Stand at side B and look through the milky fluid toward side A.

**6.** What do you see if you look through the container from side C to side D?

**7.** Record what happened in your log book.

 **CONCLUSION:** Record in your log book.

## BRAIN BUSTER

What happens if you use colored filters (or colored cellophane) on the flashlight?

## TIPS TO MAKE IT TOPS!

You have all the colors of the rainbow at your fingertips. Use them to make the backdrop of your display a beautiful sunset. Or use paper in rainbow colors to accent your texts.

Create a drawing showing how the sun's rays travel through less atmosphere when the sun is overhead than when it is low on the horizon.

## Under Pressure
### (Experiment)

## PROBLEM: Does water pressure increase with depth?

### KEY WORDS

force,

water pressure

**RESEARCH:** Go to the library and research water pressure. Visit a local scuba-diving shop and ask how water pressure affects the human body. Interview an engineer and ask how water pressure is used to move heavy objects.

**HYPOTHESIS:** Record in your log book.

### MATERIALS

- nail

- empty 1-quart milk carton

- ruler

- masking tape

- 10 inches of heavy string

- scissors

- aluminum foil

- double-sided tape

- several small paper cups (bathroom dispenser size)

- marker

## PUT IT TO A TEST

**1.** Have an adult use the nail to punch three holes of the same diameter in one side of the milk carton. Make the first hole 1 inch from the bottom and ½ inch from the right edge. Make the second hole 2 inches above that and ½ inch to the left of the first hole. Make the third hole 2 inches higher and ½ inch to the left of the second hole. Place a strip of masking tape over each hole.

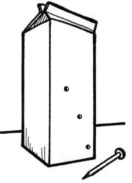

**2.** Open the top of the carton all the way. Ask an adult helper to punch two more holes, one on either side of the opening about an inch from the rim. Tie one end of the string

through one hole and the opposite end of the string through the other hole. Loop the string handle over the faucet of your kitchen sink so that the water runs into the carton.

**3.** Create a target in the bottom of the sink with a strip of aluminum foil 1 foot across. Place three long strips of double-sided tape over this foil. Place the tape strips ½ inch apart, just the way the holes in the carton are ½ inch apart.

**4.** Press the small paper cups side by side onto the strips, creating three rows. With a marking pen, label the cups in each row, with #1 being closest to the milk carton.

**5.** Fill the carton with water. Remove the strips of masking tape to allow water to shoot out of the holes. Keep filling the carton slowly from the faucet. Which stream shoots farthest? Check the cups to see which ones have water in them.

**6.** Record what happened in your log book.

 **CONCLUSION:** Record in your log book.

# BRAIN BUSTER

Would your experiment turn out differently if you used salt water? Would the water pressure be greater on a scuba diver in the salt water of an ocean or in the freshwater of a lake?

# Water, Water Everywhere
### (Demonstration)

## PURPOSE: To show how rain forms.

### KEY WORDS

atmosphere,
clouds,
evaporation,
precipitation,
rain, steam,
water vapor

### MATERIALS

- water
- tea kettle
- stove or hot plate
- metal pie plate
- ice
- oven mitt

**RESEARCH:** Go to the library and research the water cycle. Interview a meteorologist about how rain forms. Also ask about cloud-seeding and how it works.

## PUT IT TO A TEST

**1.** Put about 3 inches of water in the tea kettle. Place it on a hot plate or a stove. Have an adult helper turn on the heat.

**2.** When the kettle begins to boil, fill the pie plate with ice.

**3.** Wearing a protective oven mitt, hold the plate about 6 inches above the kettle. Observe what

happens as the water boils. What do you see? What happens to the bottom of the pie plate?

**4.** Record what happened in your log book.

**CONCLUSION:** Record in your log book.

# RAIN BUSTER

Here is a way to measure the size of raindrops on a rainy day. Fill a baking pan with about ½ inch of flour. Run outside in the rain for a few seconds and catch a few drops of rain in the pan. The drops will form little balls of dough. Let the balls dry for a couple of hours until they harden. Strain the hardened drops from the flour and measure the diameter of each one. They will be approximately the size of the raindrops that made them.

## TiPS TO MAKE iT TOPS!

To jazz up your demonstration, gather photographs and cut out pictures from magazines of various kinds of precipitation, such as rain, hail, snow, sleet, dew, and fog.

# Rock On
### (Demonstration/Collection)

## PURPOSE: To show how the rock cycle works.

### KEY WORDS

erosion,
igneous rock,
metamorphic
rock, molten,
sedimentary
rock, weathering

### MATERIALS

- your rock collection
- rock and mineral identification guide
- poster board
- marking pens
- sturdy pegboard for display
- small hooks and shelves for use with the pegboard
- labels
- index cards
- flip card file

**RESEARCH:** Visit a local rock store and check out samples of different kinds of rock. Go to the library and research rocks and minerals. Obtain a field guide to rocks and minerals. Have an adult accompany you on a collecting trip, then try to identify what you find. Interview a geologist from a local university.

## PUT IT TO A TEST

**1.** Prepare a poster to show the rock cycle. Draw a large circle using arrows as shown on the next page. At the bottom of the circle, write IGNEOUS ROCK. Below this heading, write something like: "Rocks carried deep within the earth become molten. Molten rock travels to the surface and hardens." (Use your own words.) Your arrow should point toward the text described in the next step.

**2.** Keep moving left around the circle. A third of the way around the circle, write SEDIMENTARY ROCK. Include a description. Your arrow should point toward the text described in the next step.

**3.** Two-thirds of the way around the circle, write METAMORPHIC ROCK. Include a description. Your arrow should point toward the text below IGNEOUS ROCK. The circle is now complete.

**4.** Insert shelves on the pegboard to display your rock collection. Keep in mind that if the collection is too heavy, your display could tip over!

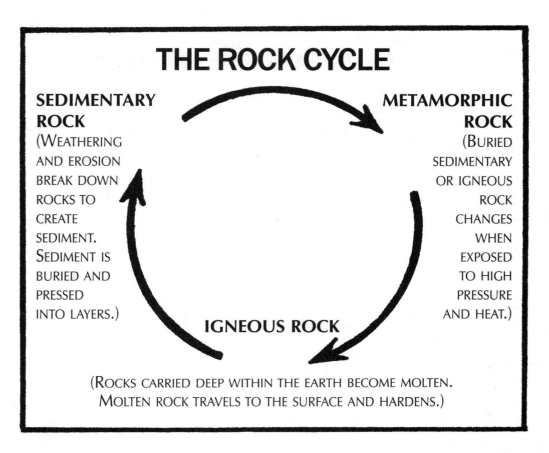

# THE ROCK CYCLE

**SEDIMENTARY ROCK**
(WEATHERING AND EROSION BREAK DOWN ROCKS TO CREATE SEDIMENT. SEDIMENT IS BURIED AND PRESSED INTO LAYERS.)

**METAMORPHIC ROCK**
(BURIED SEDIMENTARY OR IGNEOUS ROCK CHANGES WHEN EXPOSED TO HIGH PRESSURE AND HEAT.)

**IGNEOUS ROCK**
(ROCKS CARRIED DEEP WITHIN THE EARTH BECOME MOLTEN. MOLTEN ROCK TRAVELS TO THE SURFACE AND HARDENS.)

**5.** Label each rock with a number. Provide a flip card file that corresponds with the numbers and gives detailed information about each rock. For instance, you might include such information as the type of rock it is, where it was found, and its identifying features.

NUMBER: 204
TYPE OF ROCK: metamorphic
NAME OF ROCK: graphite
WHERE FOUND: Bedham Wood Road, Pine City, New York
FEATURES: black or dark gray, greasy feel, flexible, six-sided crystals

## TIPS TO MAKE IT TOPS!

Use the space in front of your display to exhibit large rocks and the tools you use when rock hunting. When preparing your poster of the rock cycle, include illustrations or maps of where the sample rocks might be found.

# The Root of the Problem

**(Experiment)**

## PROBLEM: Do plants help prevent soil erosion?

### KEY WORDS

conservation, erosion, root, soil

### MATERIALS

- two 1-inch-deep baking trays
- measuring cup
- small pebbles or gravel
- potting soil
- rye grass seed
- misting bottle filled with water
- newspaper
- two 1-inch blocks of wood
- aluminum foil
- watering can

**RESEARCH:** Go to the library and research the parts of a plant, particularly the roots. Interview a local farmer about methods of soil conservation. Interview someone from your fire department and ask how a fire affects hillsides burning away plant material. Write to the Department of Forestry for information on how plants help to preserve soil.

**HYPOTHESIS:** Record in your log book.

## PUT IT TO A TEST

**1.** Prepare both baking trays at the same time. Spread out 1 cup of small pebbles or gravel in each, then add enough potting soil to cover the pebbles and fill each tray to the rim. Don't add seeds to tray #1.

**2.** Sprinkle ½ cup of grass seed evenly over tray #2. Cover the seed lightly with soil, then mist it until it's moist. Place the tray in a warm, sunny spot. Allow the seed to sprout and grow until the grass is 1 inch high (1 or 2 weeks). Be sure to keep the soil moist but not soaked.

**3.** Once you have a growth of grass, spread newspaper on a flat surface outside. Place the two trays side by side on the newspaper. Put a wood block under each tray so they are tilted upward. Place aluminum foil under the bottom edge of the trays, and crimp the edges to catch any soil that washes away during the experiment.

**4.** With the watering can, pour ½ cup of water over each tray. Check the aluminum foil at the bottom. Which tray loses the most soil? Repeat the watering procedure once a day for one week. How does each tray look at the end of the experiment?*

**5.** Record what happened in your log book.

**CONCLUSION:** Record in your log book.

*Be sure to label your experiment, even at home. One sad science fair participant reportedly worked very hard on her plant experiment for a week. On coming home from school one day, she found that the gardener had thrown it away!

## TIPS TO MAKE iT TOPS!

Grow a sample of grass in a glass jar. Display the jar in your exhibit section to show how roots spread throughout the soil and help to hold it in place.

# Shake and Quake

### (Demonstration)

## PURPOSE: To show how a seismograph works.

### KEY WORDS

earthquake,
epicenter,
seismic wave,
seismograph

### MATERIALS

- cardboard box (approximately 12 inches square) with an open top

- marking pen

- knife

- roll of adding machine paper

- small paper cup

- 23-inch-long piece of string

- adhesive tape

- scissors

- ruler

- pebbles or gravel

### RESEARCH:

Call a local university and interview a geologist. Go to the library and research earthquakes. Interview an engineer about how buildings are constructed to survive earthquakes.

## PUT IT TO A TEST

**1.** Label the four sides of the box A, B, C, and D. Side A should be opposite side C, and side B should be opposite side D. Label the bottom E.

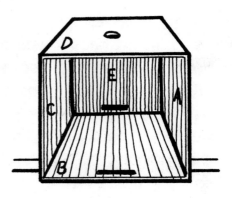

**2.** Have an adult punch a small hole in the center of side D. Then put the box at the edge of a table, side B down, so the opening is in the front and the side D hole is on top. Ask your adult helper to cut two slits large enough for the adding machine paper to slide through. Make slit #1 near the open edge of side B as shown. Make slit #2 at the lower edge of side E also as shown.

**3.** Place the adding machine paper behind the box. Slip the loose

end through slit #2 into the box, then through slit #1 out of the box. Let the loose end hang freely in front of side B.

**4.** Punch two holes just below the rim of the paper cup. In the bottom of the cup, punch one more hole just large enough for the marking pen to fit in snugly.

**5.** Tie the string through the holes at the cup rim as shown. Pinch the string together and pass the center of it through the hole in side D of the box. Tape it in place. The opening of the cup should be facing upward. The bottom of the cup should be about an inch above the bottom of the box.

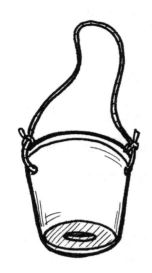

**6.** Push the uncapped marking pen through the bottom of the cup until the tip just touches the adding machine tape. Fill the cup with pebbles or gravel.

**7.** As you shake the box from side to side, hold the loose end of the adding machine tape and pull it toward you slowly. You don't need to lift the box. Just put one hand on the side and shake it.

**8.** Record what happened in your log book.

 **CONCLUSION:** Record in your log book.

## BRAIN BUSTER

The apparatus you have made is a horizontal seismograph. It records movement from side to side. What if the movement is up and down? Design a piece of equipment that could record such movement.

# Natural Beauty
### (Demonstration)

## PURPOSE: To show how crystals grow.

### KEY WORDS
crystal, gem, geode, solution

### MATERIALS

- water
- measuring cup and spoons
- stainless steel pan
- three glass jars
- cream of tartar
- mixing spoon
- labels
- marking pen
- camera
- alum (found at grocery stores)
- sugar
- scissors
- lightweight string
- paper clip
- pencil
- paper towel
- magnifying glass

### RESEARCH:

Go to the library and research crystals. Look through the phone book for a local geologist or gemologist to interview. Have an adult take you to a local rock shop or a county fair that has a rock display. Check out exhibits at a local museum.

## PUT IT TO A TEST

**1.** Pour ½ cup of water into the pan. Ask an adult to heat it almost to boiling, then carefully remove it from the heat.

**2.** Rinse one of the glass jars with hot water to warm it up. Have an adult helper fill the jar with the heated water from the pan.

**3.** Add ¼ teaspoon of cream of tartar and stir until dissolved. Label the jar, and place it in a protected spot where it won't be disturbed.

**4.** Check the jar in 2 hours to see if any crystals have begun to form. Let the solution sit undisturbed for a week. Without disturbing the jar, check it every day and draw or photograph what you see.

**5.** For giant crystals, follow the same directions, but this time use 3 tablespoons of alum. Label the jar, and leave the solution undisturbed for 2 weeks. Check it every day for signs of crystals.

**6.** For crystals you can eat, again follow the same directions, but this time add ½ cup of sugar. After making the solution, cut a piece of string that's an inch shorter than the height of the jar. Tie a paper clip to one end of the string. Tie the other end of the string to a pencil. Place the pencil across the rim of the jar so that the string and paper clip dangle into the solution as shown. Cover loosely with a paper towel. Leave the solution undisturbed for 2 to 4 days, then check to see if crystals have formed on the string.

**7.** Using a magnifying glass, compare the different crystal formations you make. Draw the crystals and record the information in your log book.

 **CONCLUSION:** Record in your log book.

### TiPS TO MAKE iT TOPS!

To add extra interest, create a geode as described above, but coil a small piece of clean, lightweight string inside each eggshell before adding the sugar solution. Crystals will form all along the string.

# BRAIN BUSTER

Have you ever seen a geode? It is a hollow rock with crystals inside. To create your own geode, carefully crack open a few eggs. Save the contents to use for cooking or baking. Rinse out the eggshell halves that have no cracks, and place them in an egg carton. Pour the cooled sugar solution into the eggshell halves, filling them almost to the top. (If you like, add food coloring to the solution before pouring it.) Leave the egg carton undisturbed for 2 weeks.

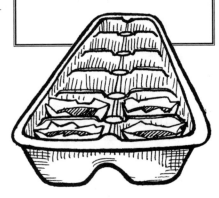

# Solar Still
### (Demonstration)

## PURPOSE: To use solar power to distill freshwater.

## KEY WORDS

condensation,
distillation,
evaporation,
solar energy,
solar power

## MATERIALS

- old aluminum pan (9 inches by 13 inches and 4 or 5 inches deep)
- flat black paint
- paintbrush
- small glass bowl
- measuring cup
- salt
- hot tap water
- three small rocks
- clear plastic wrap
- large rubber band

**RESEARCH:** Go to the library and look up distillation. Ask an adult to take you to a backpacking or camping store, where you can interview a person about survival information. Call a local college or university and talk to someone in the atmospheric sciences department about solar power. Visit a planetarium that has a telescope with a solar filter.

## PUT IT TO A TEST

**1.** Paint the inside and outside of the aluminum pan with black paint. Let it dry.

**2.** In the small bowl, mix ¼ cup of salt and 2 cups of hot tap water. Stir until the salt dissolves.

**3.** Pour the salt water into the aluminum pan. Place the bowl in the center of the pan. (The bowl needs to be at least 1 inch shorter than the pan.) If the bowl floats, put a clean rock in it to hold it in place.

**4.** Cover the whole pan with plastic wrap and secure it with a rubber band.

**5.** Carefully place a small rock or two on the plastic wrap, creating a dip. The dip should be centered over the bowl, and about ½ to 1 inch above the bowl's rim.

**6.** Place the pan in bright sunlight for several hours.

**7.** Record what happened in your log book.

 **CONCLUSION:** Record in your log book.

Does it make a difference if you cover the pan with foil instead of plastic wrap? Would you collect more water if the pan were painted white instead of black? Why can't people survive by drinking salt water?

# How They Work

Below, you'll find explanations of how each of the science fair projects work. Did you come to the same conclusions?

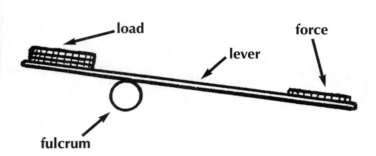

**BALANCING ACT (Page 18):** A lever can be any board, stick, or bar (or in this case, a tongue depressor or a craft stick) that turns freely on a fixed point of support. The fixed point is called the fulcrum. In this experiment, the pencil is your fulcrum. The closer the fulcrum is to the load, the less force is needed to lift it. A lever is known as a simple machine.

**RUN IT UP THE FLAGPOLE (Page 20):** With a fixed pulley, the pulley wheels (the spools) do not move with the load (in this case, the flag). A fixed pulley is a simple machine made up of one or more wheels, with a string (or a rope, a belt, or a chain) moving around them. When you pull down on the rope, the load goes up. It is easier to pull a weight down than to lift a load straight up, so a fixed pulley helps you do the work with less effort. Just think how tired you would be if you had to scurry up the flagpole every morning and evening!

**MIX IT UP (Page 22):** The sugar and salt seem to disappear when they dissolve in the hot water. They are still there. You cannot see them, but you can taste them. The tiny particles (called molecules) that make up the sugar and salt move between the tiny molecules that make up the water. They form what is known as a *solution*. The salt and sugar are called *solutes*. The water is the *solvent*. If you allow the water to evaporate, what remains of the salt and sugar is called a *residue*.

The molecules of the nutmeg and allspice do *not* interact with the molecules in the water, and they do not change how they look. Instead of dissolving, they are *suspended* in the water. The water and spices are called a *suspension*. Over time, some of the spice slowly settles to the bottom of the jars. The settled layer is known as *sediment*. If you stir the jar, the spices and water form a suspension again.

**ANCHORS AWAY (Page 24):** At its surface, water forms a sort of stretchy "skin." This is called *surface tension.* Surface tension is what causes water to form into droplets. It can also slow down a boat cruising across water. Certain hull designs are more effective than others in cutting surface tension, which is strongest at the front of the boat.

**liquid is pulled into the core**

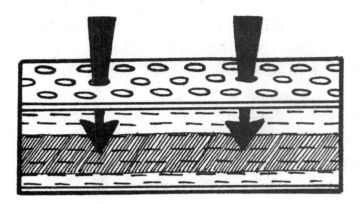

**DUELING DIAPERS (Page 26):** The pads in baby diapers are made up of lots of tiny fibers that absorb (soak up) liquid through a process called *capillary action.* The fibers have spaces between them. The water slowly creeps into these spaces. Once the spaces are all filled, the diaper leaks.

**PUCKER POWER! (Page 28):** The copper pennies and aluminum work together to create electricity when exposed to an acid-soaked (lemon juice) disc of blotting paper. A chemical reaction between the metals and the acid causes tiny particles in the materials, called electrons, to gather at one end of the battery. The electrons have a negative (–) charge. Protons gather at the other end. The protons have a positive (+) charge. This charge difference from one end of the battery to the other creates an electrical *current.* When the current passes through the LED, the LED lights up.

**HELLO! HELLO! HELLO! (Page 30):** An echo is formed when sound waves are bounced, or reflected, off a hard surface. In your experiment, sound waves bounce off the book cover toward you. The "ear tube" boosts the sound. Soft surfaces, such as a towel or a sweater, soak up (absorb) sound waves.

**sound waves bouncing off**

**sound waves being absorbed**

**MAGNETIC ATTRACTION (Page 32):** Many materials contain tiny bits of nonmagnetized iron. Each bit of iron has its own north pole and south pole, but the poles are scattered in all different directions. When you stroke a bar magnet across test materials that contain iron, you force the like poles to all line up in the same direction. With the poles now lined up, the material becomes magnetized.

**GOOD CONDUCT (Page 34):** Electricity does not flow through all materials. Those that it will flow through are called conductors. Those that it will not flow through are called insulators. When you touch the two aluminum balls to a conductor, you complete the circuit and the bulb will light.

**HEAT STROKE (Page 36):** The sun's rays heat the earth. Light-colored materials reflect the sun's rays, so the inside of the white-painted cup stays cooler than the other cups. Dark-colored materials absorb rays quickly, so the insides of the green and black cups warm up. Which color absorbs the sun's rays quickest of all?

**THIS EXPERIMENT STINKS! (Page 38):** Acids usually taste sour and can eat away other substances. Many foods contain weak acids. The cabbage water (INDICATOR) reacts with the acid in food in a chemical reaction. This reaction changes the color of the INDICATOR from purple-blue (no reaction) to pink or red (meaning acid is present), or a greenish-blue if a base is present.

A base is a chemical substance that reacts with an acid to form a salt. Bases, such as baking soda, taste bitter. Your INDICATOR turns blue-green in the presence of a base. Try your INDICATOR on baking soda, then other materials.

**SKIN DEEP (Page 40):** The shortening you use in this experiment is a kind of fat similar to the fat in a human body. Fat acts as an insulator to slow the rate at which heat leaves the body. All mammals, including humans, have a layer of fat beneath their skin. The fat layer helps to keep them warm when it is cold outside. Even so, over time, body heat will eventually be lost.

**WELCOME TO THE NEIGHBORHOOD (Page 42):** Ecology is the study of how living things interact with their home, or *environment*. If you live in the country, the living things that make up the local environment might be trees, flowering plants, rabbits, deer, and plenty of insects and birds. The nonliving things include soil, water, and air. All living and nonliving things that exist in an environment make up what scientists call an *ecosystem.*

**NATURE'S LANDFILL** (Page 45): Some items break down, or decompose, faster than others due to bacteria and enzymes. Food and plant materials usually rot away very quickly. Under the right conditions such as those in your experiment, paper products also decompose fairly quickly. (Remember that paper is made from wood, a plant product.) Plastic items can take hundreds of years to break down. That means the plastic things you throw away today might still be around for your great-great-great-great-great-great-great-great-grandchildren to deal with.

**TAKE A LITTLE OFF THE TOP** (Page 48): Young plants usually have a growth point—the point from which new leaves are produced. If you cut away the growth point, the plant will die. The growth point of rye grass is at soil level. That is why it continues to grow even after it's cut with a lawn mower or chewed on by a hungry plant eater.

**YOU'VE BEEN FINGERED** (Page 50): Your skin has tiny openings called pores. Oil and perspiration flow up through the pores to the skin surface. When you touch something, the ridges (prints) on your hands leave deposits of oil and perspiration. Fine powder sticks to these deposits, revealing the fingerprints of the person who left them.

Your fingerprints are unique. They form two months before you are even born, and no one in the world has the same print pattern as you do!

**RED SKY AT NIGHT** (Page 52): The sunlight we see is made up of different colors that move at slightly different speeds. As sunlight passes through the atmosphere, it hits particles that break it up and scatter the different colors. During the day, the rays of the sun are direct. You

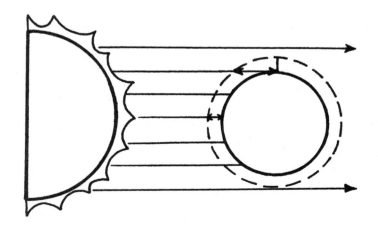

see the scattered blue light, so the sky appears blue. When the sun is low on the horizon, the rays strike the atmosphere at a low angle. Sunlight must pass through more atmosphere, which means it is scattered more. Only red light reaches your eye, so the evening sky appears red.

When the light from the flashlight passes through the liquid in the container, it strikes tiny milk particles that break it up into different colors and scatter it. When you look through the container from side B to side A, the light rays you see have been scattered a lot. You see only the longer, slower red light rays. When you look through the container from side C to side D, the light rays are less scattered, and the milk has a bluish tint.

**UNDER PRESSURE** (Page 54): Pressure is a force that one material exerts on another. The water at the surface of the milk carton exerts pressure on the water below it. The pressure of water increases with depth because of the weight of water pushing down from above. That's why the water from the lowest hole shoots out farthest.

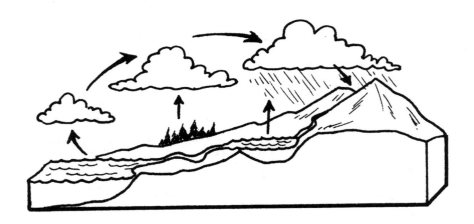

**WATER, WATER EVERYWHERE** (Page 56): In this experiment, the water boils, turning to steam (or water vapor). The water vapor condenses on the bottom of the cold pie plate and forms drops that "rain" down. One of the biggest recycling projects on earth is nature's own water cycle. Water is always evaporating from the oceans and other bodies of water. The water vapor rises high into the atmosphere, where it cools, condenses, and forms clouds. The condensed water eventually falls back to earth again as rain.

**ROCK ON** (Page 58): No explanation needed.

**THE ROOT OF THE PROBLEM** (Page 60): Soil is a layer of loose material that blankets much of the surface layer of the earth's dry land. Soil is made up of particles of rocks, minerals, and other materials, such as rotted leaves. Most plants cannot grow without soil. Loose soil can be washed away by water, or blown away by wind. Plants help to shield the soil from wind, and plant roots hold the soil in place.

**SHAKE AND QUAKE** (Page 62): The cup stays steady while the paper underneath it moves, so the pen draws a zigzag line on the paper. The cup stays in place due to *inertia,* the tendency an object has to stay in the same state (at rest or moving) unless a force changes that state. A real seismograph works the same way.

**NATURAL BEAUTY** (Page 64): A crystal is a solid substance with flat surfaces. The surfaces always meet at the same angle. Almost all minerals are made up of crystals. Crystals can form from a solid, a liquid, or a gas. The atoms that make up a particular crystal arrange themselves, layer after layer, in a pattern. As more material is added to the outer surface, the crystal grows. There are seven crystal categories, or systems, based on shape. Look up crystal systems in the encyclopedia to see how each is shaped.

**SOLAR STILL** (Page 66): In this demonstration, you use the sun's energy to separate the water from the salt. The sun heats up the salt water. The water gradually *evaporates* into water vapor. The vapor then *condenses* onto the plastic wrap. The salt is left behind in the pan. The dip in the plastic channels the water down into the bowl, where it collects as distilled water. If you taste it, you will discover that it is no longer salty.

Some arid desert countries use solar energy to obtain drinkable distilled water from sea water.

# Make It Yourself!

Models and handmade displays can make your science project stand out. The recipes below might come in handy when you're designing your presentation.

## PAPIER-MÂCHÉ MODEL

**1.** Tear old newspapers into 1-inch-wide strips.

**2.** Make a form for your model by molding recycled aluminum foil or crumpled paper into the shape you want. You can also use an inflated balloon. Cover the form with masking tape.

**3.** Mix ½ cup of flour and 2 teaspoons of salt in a bowl. Add 1 cup of warm water. Mix this paste with your hands until it is thick and creamy.

**4.** To finish the model, dip newspaper strips into the paste and hold them over the bowl until they stop dripping. Cover the form with overlapping strips of paper. Continue until it is covered with two or three layers. Finish with a final layer of blank paper strips.

**5.** Give your model several days to dry completely. Drying it on a rack will allow air to circulate and prevent mold from forming. After it's dry, it's ready to paint!

# BAKER'S CLAY

**1.** In a bowl, mix together 1½ cups of flour and ½ cup of cornstarch. In a separate bowl, mix ¼ cup of salt and ¾ cup of hot water, then add that to the flour and cornstarch. Blend with a spoon. The "clay" should stick together.

**2.** Form the clay into a ball and squash it in your hands for 5 or 10 minutes, until it's smooth and satiny. Now the clay is ready to form into any design you choose.

**3.** When you have finished your model, place it on an ungreased cookie sheet. Have an adult put it in the oven at 200°F for 1½ to 2 hours, or until it is completely hardened. A large piece may take as long as 4 hours. Don't overcook it! Once it has cooled, it's ready to paint.

# DISPLAY CARDS

**1.** Gather plant materials, such as seeds, leaves, and flowers. Arrange them as you like on a piece of colorful construction paper.

**2.** Cut a piece of clear, adhesive plastic (available at craft stores) the same size as your construction paper. Carefully place it, sticky side down, over your arrangement. Press the final product between newspaper and under a heavy book for several days before displaying.

# Resources

The following are good sources of information on various scientific subjects.

## SOCIETIES AND MUSEUMS

### Chemistry

American Chemical Society
Science Office (K-8)
1155 16th Street, NW
Washington, DC 20036
phone: (800) ACS-5558
*(When the voice menu prompts you, press #*
*for the education department.)*

### Insects

Young Entomological Society
1915 Peggy Place
Lansing, MI 48910-2553
fax and phone: (517) 887-0499
E-mail: YESbugs@aol.com

### Spaceflight

NASA Goddard Space Flight Center
Educational Office
Greenbelt Road
Greenbelt, MD 20771
Attn: Code 130
phone: (301) 286-7205

### Weather

American Meteorological Society
45 Beacon Street
Boston, MA 02108
phone: (617) 227-2425

### Geology

American Geological Institute
4220 Kings Street
Alexandria, VA 22302
phone: (703) 379-2480

### Environment and Wildlife

U.S. Fish and Wildlife Service
Main Interior Building
1849 C Street, NW
Washington, DC 20240
phone: (800) 344-WILD
Web site: http:\\www.fws.gov

National Audubon Society
700 Broadway
New York, NY 10003
phone: (212) 979-3000
Web site: http:\\www.audubon.org

National Wildlife Federation
8925 Leesburg Pike
Vienna, VA 22184-0001
phone: (800) 822-9919

The Sierra Club
85 2nd Street, 2nd Floor
San Francisco, CA 94105
phone: (415) 977-5653
E-mail: information@sierraclub.org
Web site: http:\\www.sierraclub.org

**General Science**

Field Museum
Roosevelt Road at Lake Shore Drive
Chicago, IL 60605
phone: (312) 922-9410

Fort Worth Museum of Science and History
1501 Montgomery Street
Fort Worth, TX 76107
phone: (817) 732-1631

# WEB SITES

National Hurricane Center
http:\\www.nhc.noaa.gov\

National Climatic Data Center
http:\\www.ncd.noaa.gov\ncdc.html
*(Forecasts, information on the
Tropical Storm Prediction Center, and
satellite and radar imagery)*

Kennedy Space Center
http:\\www.kscvisitor.com
*(Historical spaceflight information, space
news, space theater, and more)*

# SCIENTIFIC EQUIPMENT AND SUPPLIES

*(Write or call for catalogs, books, or kits.)*

Edmund Scientific Co.
101 East Gloucester Pike
Barrington, NJ 08007
phone: (609) 573-6250

The Science Eye
18241 SW 52nd Court
Fort Lauderdale, FL 33331
phone: (954) 680-7077

Fisher Scientific Co.
4901 West Lemoyne Street
Chicago, IL 60651

Science Kit, Inc.
777 East Park Drive
Tonawanda, NY 14150

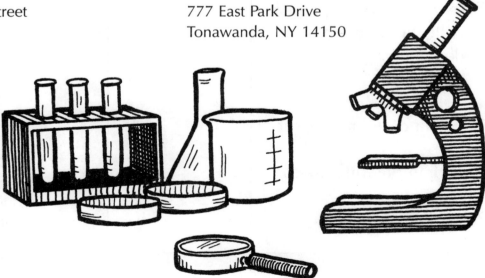

# Timetable

Use this handy timetable to help you organize your time for any science fair project. Just photocopy this page, or copy it onto a separate piece of paper.

Science fair date:

Science fair location:

| TASK | BEGIN | DEADLINE | WHEN COMPLETED |
|---|---|---|---|
| Get a copy of the rules | | | |
| Choose a topic and    write a purpose | | | |
| Do research | | | |
| Gather materials | | | |
| For experiment: <br>• Choose a problem <br>• Design an experiment <br>• Draw conclusions <br>• Perform experiment (or build model, prepare collection, design a demonstration, and so forth) | | | |
| First draft of written report | | | |
| Final draft of written report | | | |
| Prepare display | | | |

# Glossary

**absorb:** to take in. A paper towel absorbs liquids.

**angle:** the amount of space between two lines (or planes) that meet at a point or intersect each other. An angle is measured in degrees.

**biodegradable:** able to be broken down by natural agents, such as bacteria.

**chlorophyll:** a chemical in plants that is used during photosynthesis. Chlorophyll gives plants their green color.

**classify:** to arrange information in logical order.

**condensation:** the changing of a gas or a vapor into a liquid by cooling or by a lowering in pressure.

**data:** information collected during an experiment and used to find patterns or relationships.

**decompose:** to break down into a simpler form.

**evaporation:** the changing of a liquid into a gas by heating. A liquid doesn't necessarily need to reach the boiling point in order to evaporate.

**force:** the energy or power exerted on an object that makes the object move, change shape, or change direction.

**hypothesis:** an educated guess.

**inertia:** the tendency of an object to keep doing what it's doing (whether that's resting or moving) unless it is acted upon by a force.

**load:** the material being moved by a machine.

**molecule:** the smallest amount of a substance that has the properties of that substance. Molecules are made up of two or more atoms.

**observe:** to use the senses to obtain information.

**photosynthesis:** the process by which green plants produce food from carbon dioxide, water, and sunlight.

**precipitation:** any form that water takes when condensing and falling to earth. Rain, fog, snow, and dew are all forms of precipitation.

**pressure:** the amount of steady force applied to an area.

**temperature:** the measure of how hot or cold a material is.

**water vapor:** water that has become gas due to evaporation or boiling.

# Index